Listen Up!

Laboratory Exercises for Introductory Radio Astronomy With a Small Radio Telescope

Laura A. Whitlock
Louisiana State University Shreveport

Kiley Pulliam
Tulane University

iUniverse, Inc.
New York   Bloomington

Listen Up!
Laboratory Exercises for Introductory Radio Astronomy with a Small Radio Telescope

iUniverse books may be ordered through booksellers or by contacting:

iUniverse
1663 Liberty Drive
Bloomington, IN 47403
www.iuniverse.com
1-800-Authors (1-800-288-4677)

ISBN: 978-1-4401-0784-9 (pbk)
ISBN: 978-1-4401-0785-6 (ebk)

Printed in the United States of America

iUniverse rev. date: 12/3/2008

To Jocelyn Bell Burnell, for inspiring generations of women to reach for the stars.

Table of Contents

Acknowledgments

The authors wish to thank Preethi Pratap, Alan Rogers, and Phil Shute at MIT Haystack Observatory for their infinite patience and guidance, as well as all of the educators and staff at MIT Haystack who developed the laboratory foundations for using the SRT. The authors also thank Greg Andrews for his review and assistance. Lastly, this effort would not have been possible without the support of Libby W. Truelove.

Original Artwork by Aurore Simmonet.

Introduction to the Small Radio Telescope

MIT Haystack Observatory developed a small radio telescope (SRT) capable of making observations in a frequency band near 1.42 GHz. This particular frequency, which is discussed later, is of prime importance in radio astronomy. The SRT kit consists of a system of hardware and software fully integrated to allow the user easy access the radio sky. The SRT consists of a standard 7-ft diameter satellite television dish constructed from a metal mesh mounted on top of a motorized Az-El mount. The mounting arrangement allows the user to perform both total power measurements and contour maps of radio sources. Software is provided for controlling the antenna, making selection of sources, performing visual analysis of the gathered data, and exporting data for additional analysis in standard software packages such as Excel or Matlab. This inexpensive radio kit provides everything necessary to introduce users to the amazing world of radio astronomy.

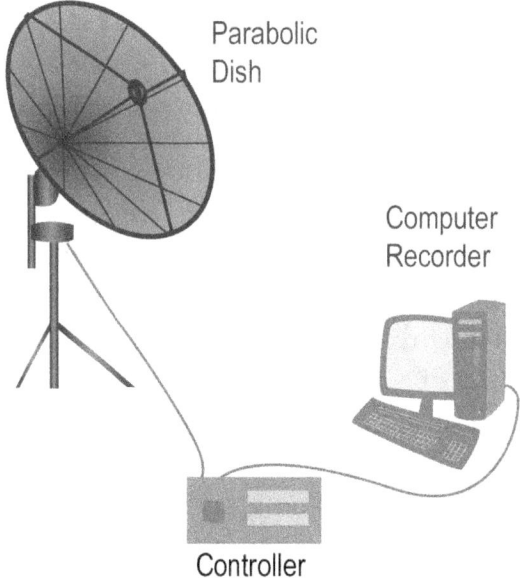

The function of each element in the SRT system is fairly straightforward. The parabolic reflector collects power from astronomical sources and reflects it to the feed at the prime focus. Behind the feed is the receiver system. The receiver amplifies the radio signal. The receiver is configured in such a way that throughout the amplification process, the signal remains directly proportional to the strength of the incoming radiation. So the resulting image or spectrum is a true representation of the emission from the astronomical source. This amplified signal is then sent to the controller for eventual output to your computer.

The SRT kit is available from Custom Astronomical Support Services, Incorporated. Information can be found at www.cassicorp.com.

The Sun as a Radio Source

Many of the labs in this book involve observations of our star, the Sun, as a radio source. Provided below is some background information about the Sun as a radio source which may help you in these labs.

As human beings, our bodies have the ability to tell us that the Sun is a source of infrared, visible, and ultraviolet radiation. We can feel it on our skin as heat (a testimony to infrared), see it with our eyes (a testimony to visible), and be burnt by it (a testimony to UV). However, our bodies cannot readily tell us that the Sun is also emitting in the radio, X-ray, and gamma ray regions of the electromagnetic spectrum. But it does, and just like it is the dominant source of visible light in our sky, it is the dominant source of radio emission in our sky. However, dominant does not have quite the same meaning for the two skies. In the visible, the Sun's light is scattered off of the particles in Earth's atmosphere giving us a bright blue sky all over, in addition to a bright discrete source in it. Radio waves are too long to be scattered by the atmospheric particles, so the Sun makes a bright discrete source in an otherwise dark radio sky.

The Sun emits radio waves through several different mechanisms: (1.) the synchrotron process, which involves high-speed electrons spiraling around magnetic fields: (2.) the thermal emission process from the hot plasma; and (3.) natural oscillations of the solar plasma itself. The thermal emission component consists of two parts, one of which is always there because the Sun is a hot ball of ionized gas. This part is often called the "quiet Sun" component. A second component of the thermal emission arises from the regions directly above sunspots. The magnitude of this component, being directly related to the number of sunspots visible on the solar surface, will then vary over the 11-year solar cycle. The other components create the "active Sun" components, and can vary on short time scales depending on solar activity such as flares and the appearance or disappearance of sunspots. These components can vary in frequency and over time in very complex ways.

The measured appearance of the Sun in the radio sky depends on the wavelength of your observation. The main source of radio emission is from electrons, and the frequency of emission depends on the depth in the solar atmosphere at which the emission takes place. This is often discussed in terms of opacity; opacity is just a measure of how much a wave gets absorbed as it travels through a medium. The bulk of the emission arises from the region where the opacity, τ, is near 1. At visible wavelengths this happens in the photosphere where the temperature is about 6000 K. At a frequency of 1.4 GHz (21-cm wavelength) the emission originates from the top of the chromosphere where the temperature is about 100,000 K. At longer wavelengths (300 cm or frequency of

0.1 GHz) the emission arises from the corona where the temperature is about 2 million K. So it should be no surprise that the apparent size of the Sun varies with different observational wavelengths or frequencies.

The Quiet Sun

Image courtesy of NRAO/AUI and Image courtesy of Stephen White, University of Maryland, and of NRAO/AUI.

The image above is from data taken by the Very Large Array radio telescope in New Mexico and shows the quiet Sun at a frequency of 4.6 GHz. The two small dark oval features in the patch of light grey show where extra strong magnetic fields exist in the Sun's atmosphere, and are associated with high temperatures. An optical image of the atmosphere under these areas would usually show sunspots. The light grey features show where the Sun's atmosphere is very dense, but cooler than it is above the sunspots. The dark patchy features are the coolest areas of the Sun in the image. The long slash across the bottom of the disk is called a "filament channel, where the Sun's atmosphere is known to be very thin.

By the way, the Small Radio Telescope you will be using does not have the resolution found in the image above. When an image of the Sun is taken with the SRT, it looks like this:

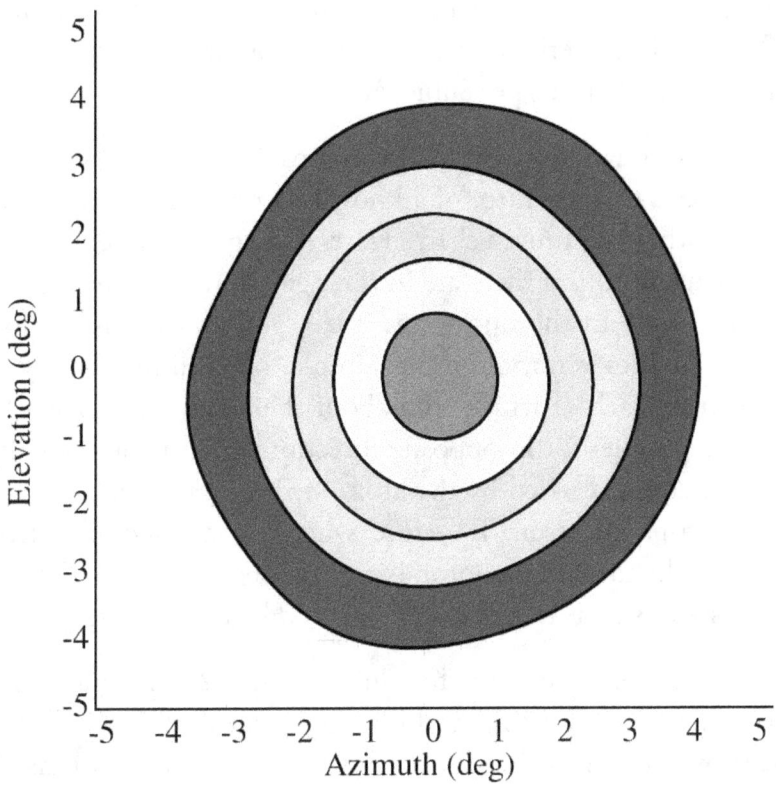

Web sites for more solar data:
http://www.sec.noaa.gov/Data/solar.html
http://ovsa.ovro.caltech.edu/
http://solar.nro.nao.ac.jp/
http://www.sao.ru/hq/sun/
http://www.spaceweather.com
http://web.haystack.mit.edu/pcr/precollegeindex.htm

The 1.42 GHz (or 21-cm Line) Emission

In astronomy, we refer to regions of space where matter is composed of neutral hydrogen atoms as H1 regions. These regions may be between stars in a galaxy, and may even form denser clouds of such cold material in certain areas. You are considered a cloud if your particle density is at least 10 atoms per cubic cm.

Hydrogen is the most abundant element in the cosmos; it makes up 80% of the universe's mass. Therefore, if it can be the source of a known spectral line, it can be an incredibly valuable tool for exploring the universe! In H1 regions, gas is extremely cold (~ 100K) and so the hydrogen atoms found there are in their electronic ground states. This means that the electron is as close to the nucleus as it can get, and it has the lowest allowed energy. However, the physics is not quite that simple. Neutral hydrogen consists of one proton and one electron, each of which spin about their individual axes. They can spin in the same direction (parallel) or in opposite directions (anti-parallel) while they are in the ground state. The energy carried by the atom in the parallel spin is greater than the energy it has in the anti-parallel spin. Therefore, when the spin state flips from parallel to anti-parallel, energy in the form of a photon is emitted. For neutral hydrogen, the energy of this photon corresponds to a frequency of 1.420 GHz, or a wavelength of 21-cm.

Besides hydrogen being everywhere in the Universe, the 21-cm line is also extremely useful as a probe because it is not impeded by interstellar dust. This allows us to probe deep into the heart of our own Milky Way Galaxy. Optical observations of our Galaxy are limited due to the interstellar dust, which absorbs visible light waves. However, this problem does not arise when making radio measurements at 21-cm. This radiation can be detected anywhere in our Galaxy, and variation in the detected frequency can be used to determine the motion of the emitting material via the Doppler shift.

Lab 1: Introduction to SRT software

Objective: To learn how to use the SRT software interface to manipulate the telescope and collect data. You will collect data from the Sun that will be used in a later Lab.

SRT Software:

1. Turn telescope on by flipping the switch on the controller (a small metal box with a switch). The red light on the controller should come on.

2. Open the SRT program by double-clicking on the SRT icon on the desktop.

3. An interface should appear on your screen which looks like the one shown here. Take some time to explore the options at the top of the screen by placing the cursor over each option along the top of the screen; a description of what each button does will appear in the lower left-hand corner. Additional information is listed below.

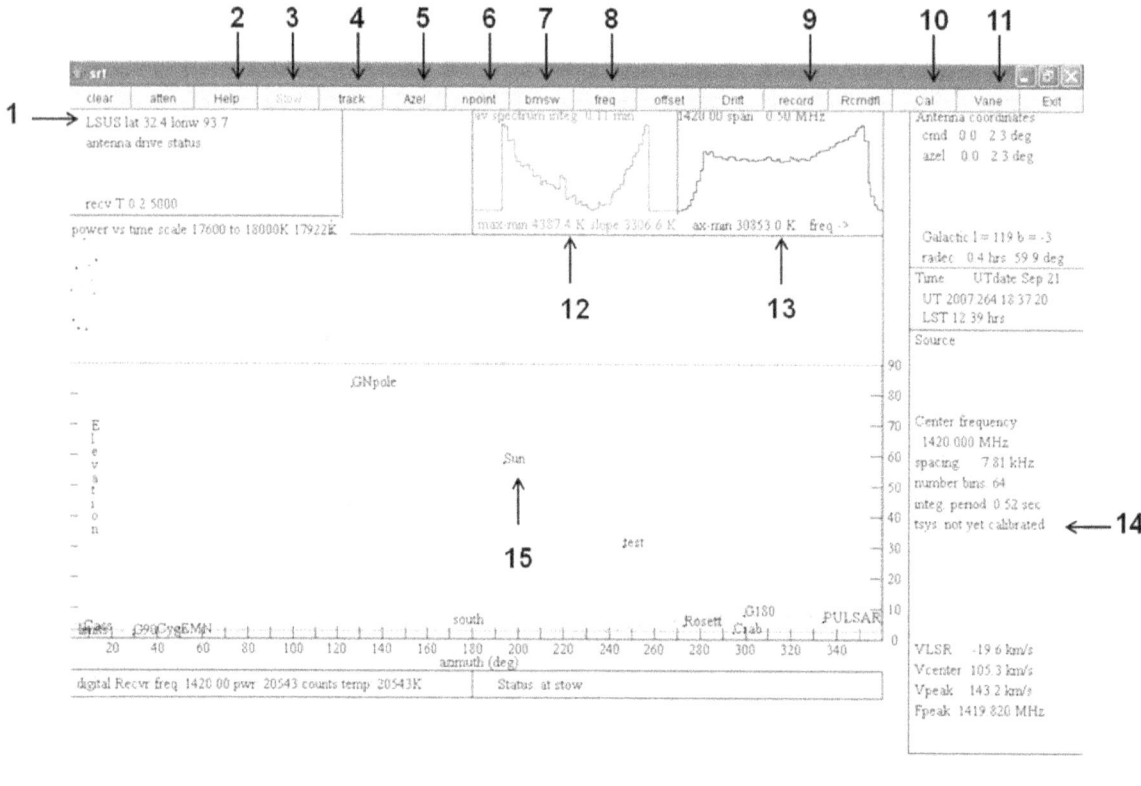

1. SRT's name and spatial location (make sure it matches your latitude and longitude; this is set in the SRT.cat file)

2. Help topics on the SRT. Be careful not to highlight or press the delete button when your cursor is over each section or it will delete the content of that section!!.

3. Pressing this button returns SRT to its stow (or rest) position

4. Pressing this button tells the SRT to remain focused on a source as the Earth rotates.

5. Pressing this button allows you to manually input the azimuth and elevation coordinates of a source you want the SRT to point at. This is necessary in order to point the SRT at an 'empty' region to calibrate the system noise (see 10).

6. Pressing this button starts one 25-point scan of the source the SRT is currently pointing at, but the source must be one of the visible icons on the interface (see 15).

7. Beamswitch

8. Pressing this button allows you to set the central frequency and frequency range modes at which the SRT's receiver looks. Averaged and instantaneous data appear in 12 and 13, respectively.

9. Pressing this button once starts data recording to an output file which, by default, is saved as YYDDDHH.rad. Pressing the button a second time stops the data recording.

10. Pressing this button begins an automatic system calibration which measures the 'noise' produced by the SRT's own components. The value obtained will display at 14 and be automatically subtracted out of any data recorded after the calibration.

11. Pressing this button allows the user to perform a manual calibration of the system; not usable in some SRTs.

12. The red plot is the accumulated spectrum with a baseline constant and slope removed in beamswitch mode. The difference in total power from the sum of all frequencies is given with error estimate.

13. The black plot is an instantaneous spectrum of the last reply from the receiver.

14. When you calibrate using the 'Cal' button, the value obtained will display here and subsequently, be subtracted from any measurements.

15. This window shows the SRT catalog of sources it recognizes and can point at during a given time. As the Earth rotates, the sources will move clockwise across this window. Only sources visible in the window can be observed, so all measurements of a source like the Sun must be made during the morning or early afternoon.

4. To have the SRT point at a source, like the Sun, simply use your cursor to click on sun; the box in the upper-lefthand corner should indicate that the SRT's drivers are moving in azimuth and elevation. Once the SRT has reached the source, red crosshairs will appear over the source on the interface. Alternatively, if you wish to point in a location where there is no icon, click on the 'Azel' button at the top of the interface and manually input the desired coordinates and press enter.

5. Try pointing at several different sources both by clicking on a source icon, and manually setting the coordinates in 'Azel'. It should be noted that if you choose to input the coordinates of a listed source, such as the Sun, the 'track' button will only track the coordinates you input, not the source you are interested in.

SRT data collection:

1. Anytime you are going to use the SRT to collect data, the telescope must be calibrated. *It is very important that you follow these directions as it will effect your data output!*

   - Select an empty region on the screen for the radio telescope to point that is between 15 and 20 degrees away the source of interest. To point the SRT use the Azel button to set the coordinates of this region and press enter.

   - When the telescope reaches this point (the crosshairs that showed the telescope's movement should now be at the coordinates you choose). If you see a lot of activity on the graphs at this location, point somewhere else since you want an area with low activity to measure the system noise of the SRT. Once you find such an area, click the Cal button at the top of the screen.

   - Look at the lower right-hand side of the screen at the value of tsys. This value represents the 'noise' of the system from the telescope's machinery and will be subtracted from any subsequent data recorded.

   - Now point the telescope at the Sun

2. Once the telescope is pointed at the source, click on the track button. It will turn green. This will ensure that as the Earth rotates, the telescope continues to point at the source.

3. Now you are ready to record data. Click the record button at the top of the screen. It will turn green. Note the file name your data is being recorded to (the information in red in the lower right-hand side of the screen).

   - Data file output is by default recorded in the format YYDDDHH.rad

- Data files can be located in C:\SRTcassi

4. Record data for one minute (one minute is sufficient for the Sun regardless of weather conditions), click the record button again. It will turn black.

5. When you are finished using the telescope, click the Stow button to return the telescope to its initial position, turn the telescope off by flipping the switch on the controller, and then you can exit the program.

Formatting Data in Excel

Note: Before data can be analyzed, it must be formatted in Microsoft Excel

1. Open the Microsoft Excel Spreadsheet program

2. Go to File and click Open

3. Open the data file of interest (C:\SRTcassi\YYDDDHH.rad)

4. When the Text Import Wizard comes up, select the Delimited option under the Original Data Type box and then click next.

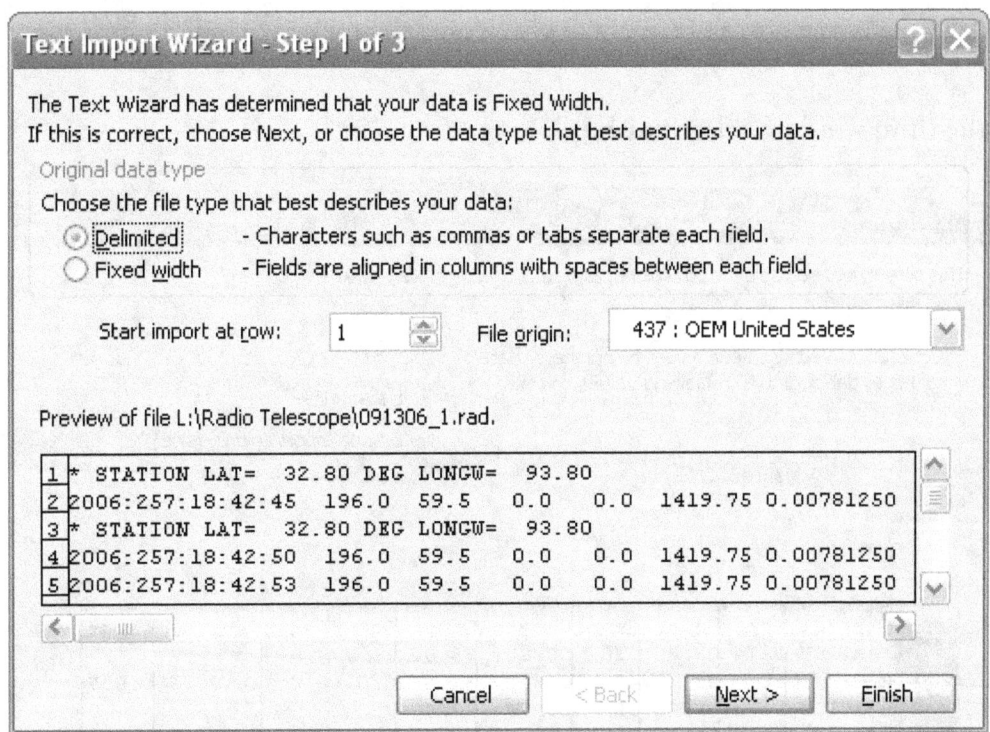

5. In the second Text Import Wizard box, check the delimiters 'space' and 'other'. Put the colon (:) in the 'other' box and then click next.

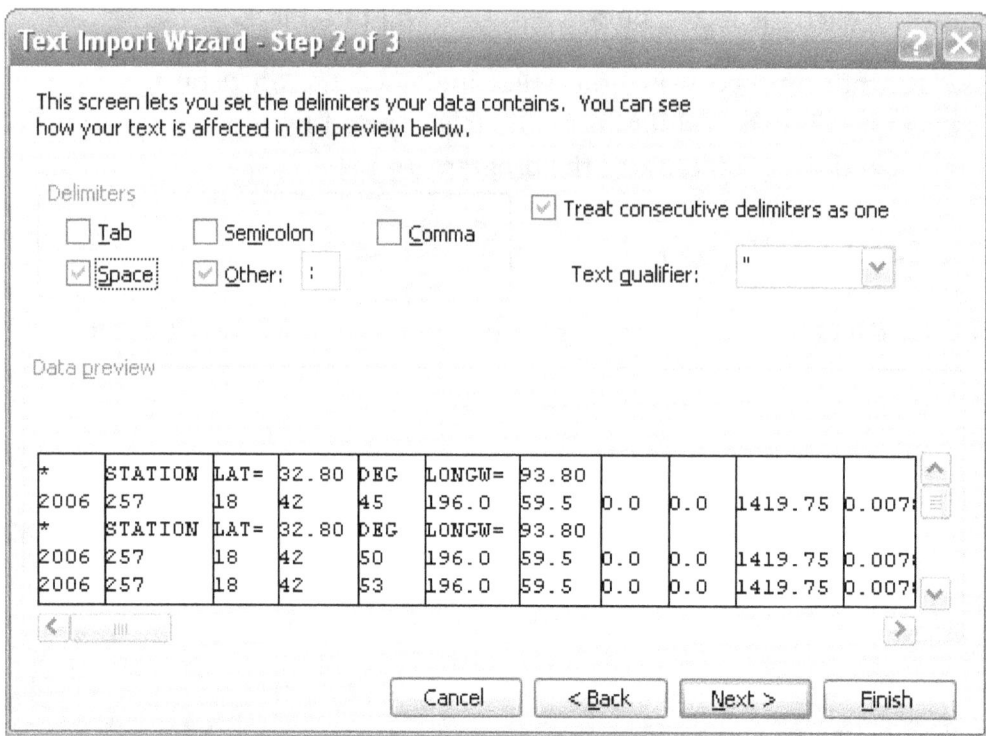

6. In the third window, select finish.

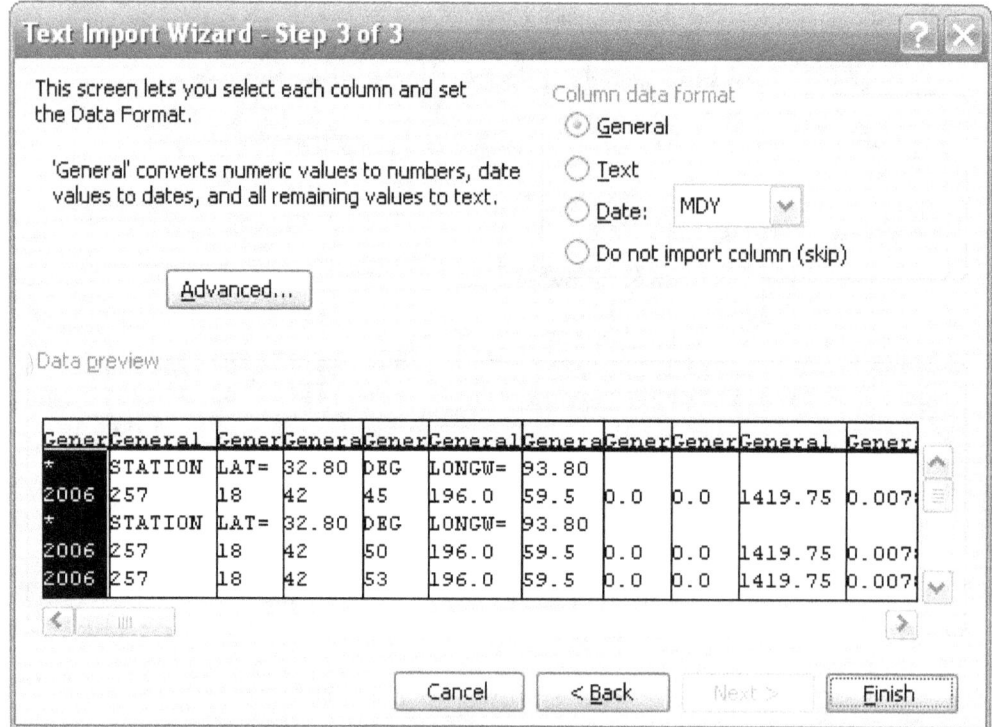

7. Your data should be organized in columns as shown below.

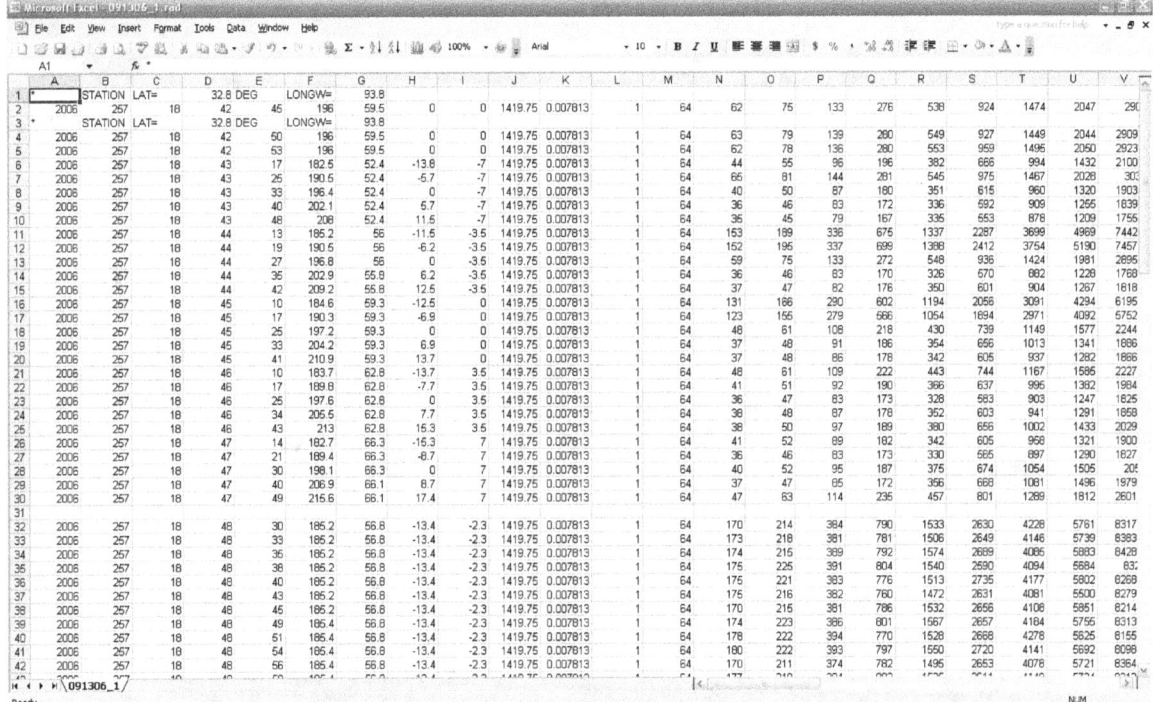

8. Use the 'Save As' option under file and save your data as a Microsoft Excel Workbook.

Questions

1. What are some possible sources of system noise (tsys) for the SRT? If you had an usually large tsys measurement, what could you do to check whether it was system noise or noise from an unlabeled source?

2. Look at the SRT help file on data output. Using this, label each column of your data.

Lab 2: Data Reduction in Excel

Purpose: To introduce students to using Excel to analyze data taken with SRT software.

Introduction: In this lab you will use the data collected from the Sun in the first lab to create plots. This can be done using Excel or MATLAB software.

Procedure:

Data Reduction using Excel

Part 1

1. Open your data file in Excel

2. Highlight *only* the first row of raw data starting with column 14 (column N)

3. Click on the chart wizard icon at the top of the Excel interface or click the Insert tab and select chart.

4. In the chart wizard, select the XY Scatter under 'chart type' and then under 'sub plots' choose one of the 'scatter with data points connected by smoothed lines' option.   Click Next.

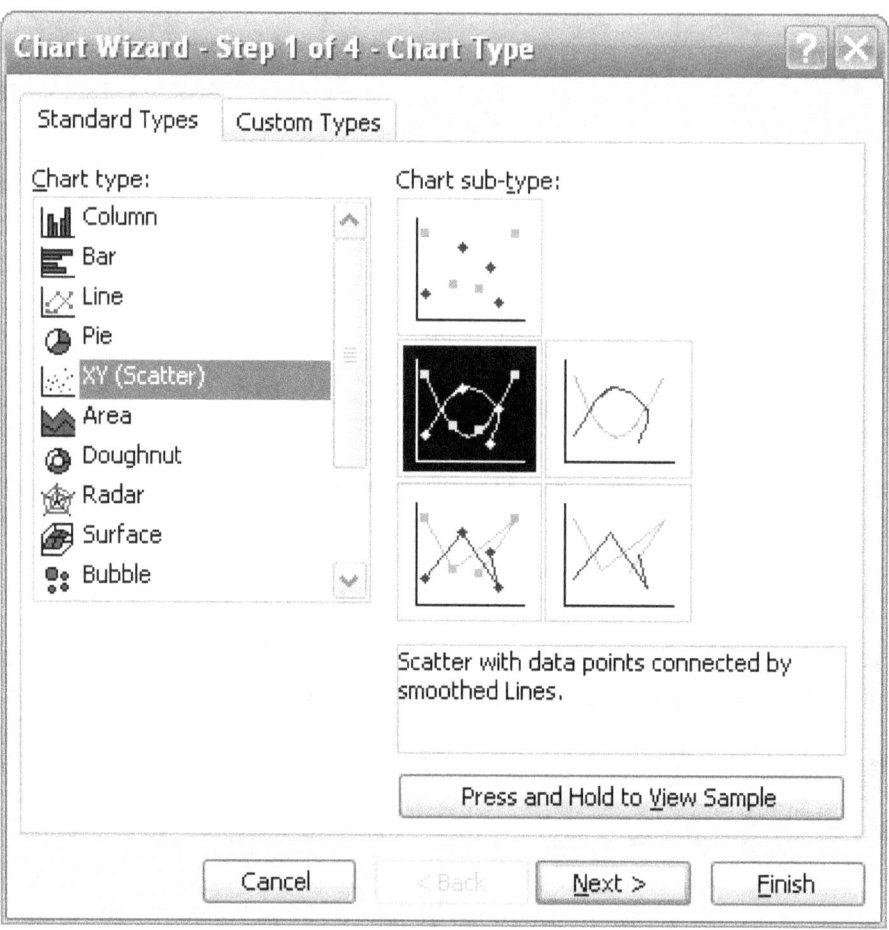

5. In Step 2, click the Series tab. Click the 'Y-Values' tab and verify that only the first row of data stating with column N is highlighted. The 'X-Values' should be empty as the number of data points will be used for the X-axis. Click Next.

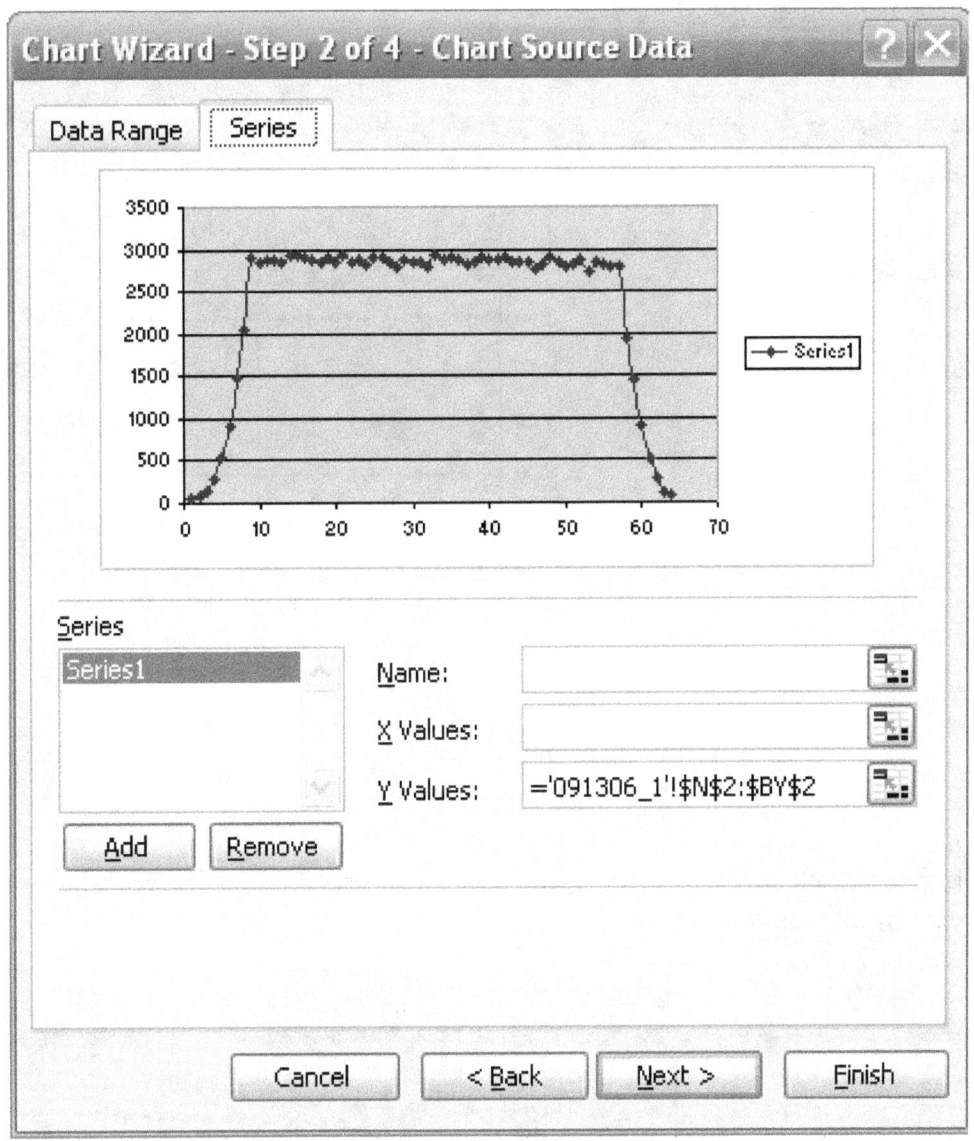

6. In step 3, label the X and Y axes (Data Bins and Intensity, respectively) and customize your graph using options under the other tabs as you see fit. Click next.

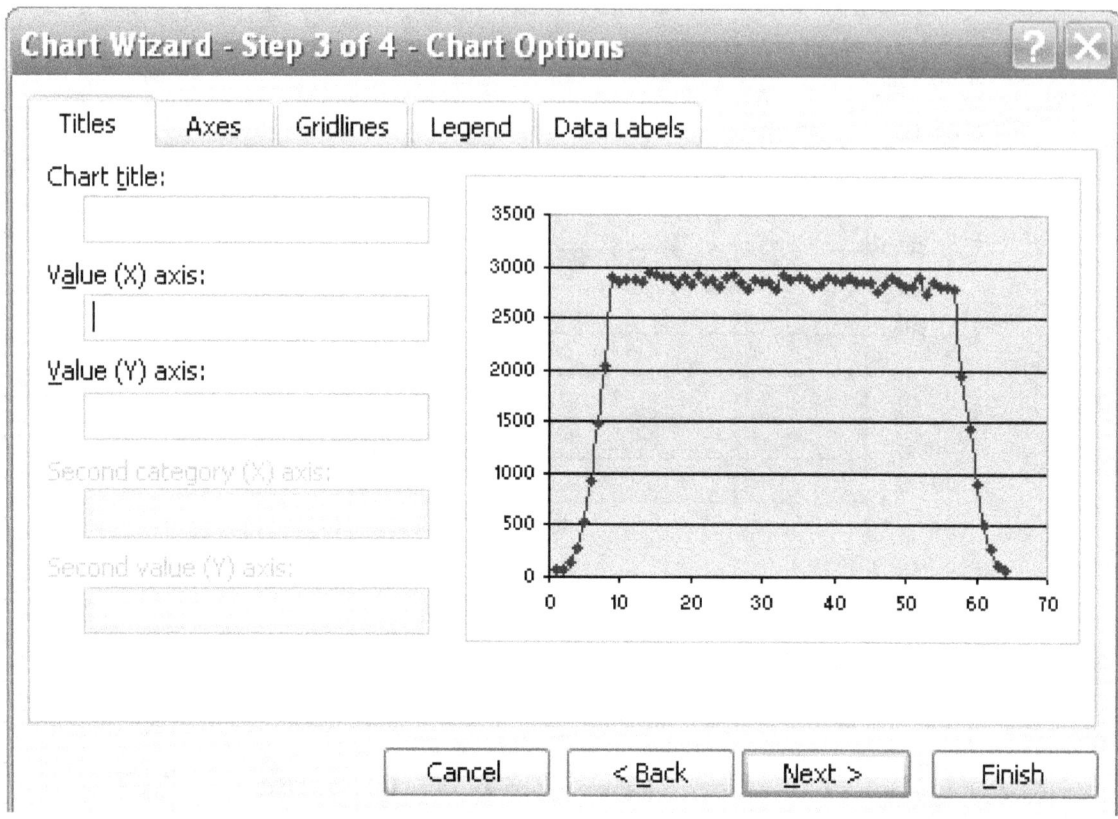

7. In Step 4, select whether you want the graph displayed in the worksheet with the data or in a separate worksheet. Click Finish.

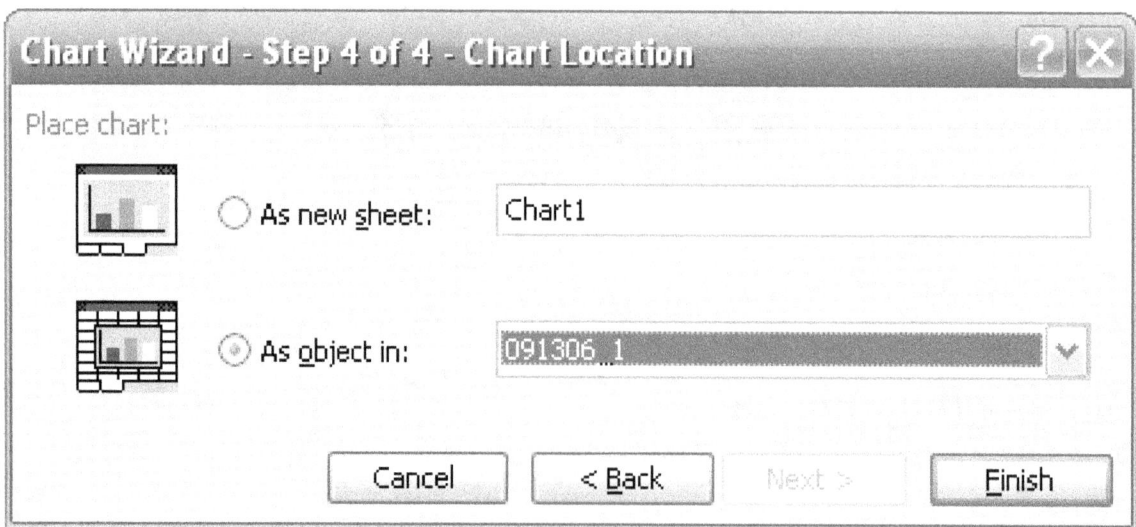

8. Your graph should look something like this:

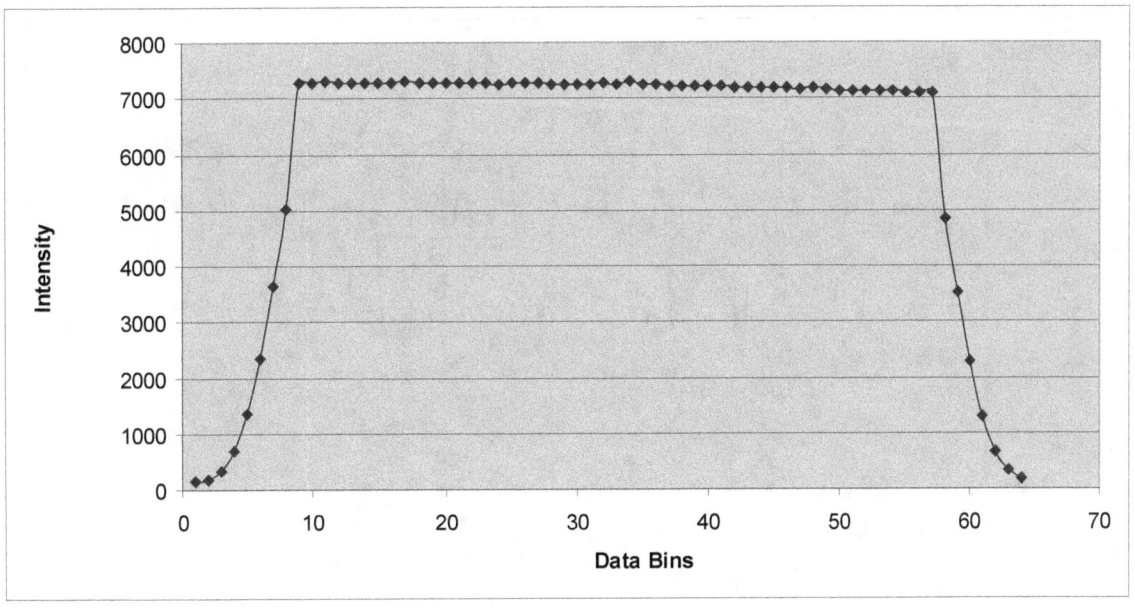

9. Note that the graph slopes up and slopes down at the extremes. This is due to limitations in response time of the receiver.

   a. To eliminate these slopes in the data, first examine your graph and place your cursor on the first data point that is not on the upward slope and write down the data point number and coordinates, then repeat this for the last data point before the graph begins to slope downwards.

   b. Now, refer back to your data table and highlight only the data points between the two points you noted in the previous step.

   c. Re-graph this edited set of data. You can either go through the graphing directions again if you wish to keep the original graph as well or you can simply right-click on the chart and select 'Source data'. Now go to the series tab and click on the 'Y-Values' and highlight your new Y-values. Then click the button on the lower-right-hand side of the 'Y-values' button and click OK.

10. Your graph should look similar to this:

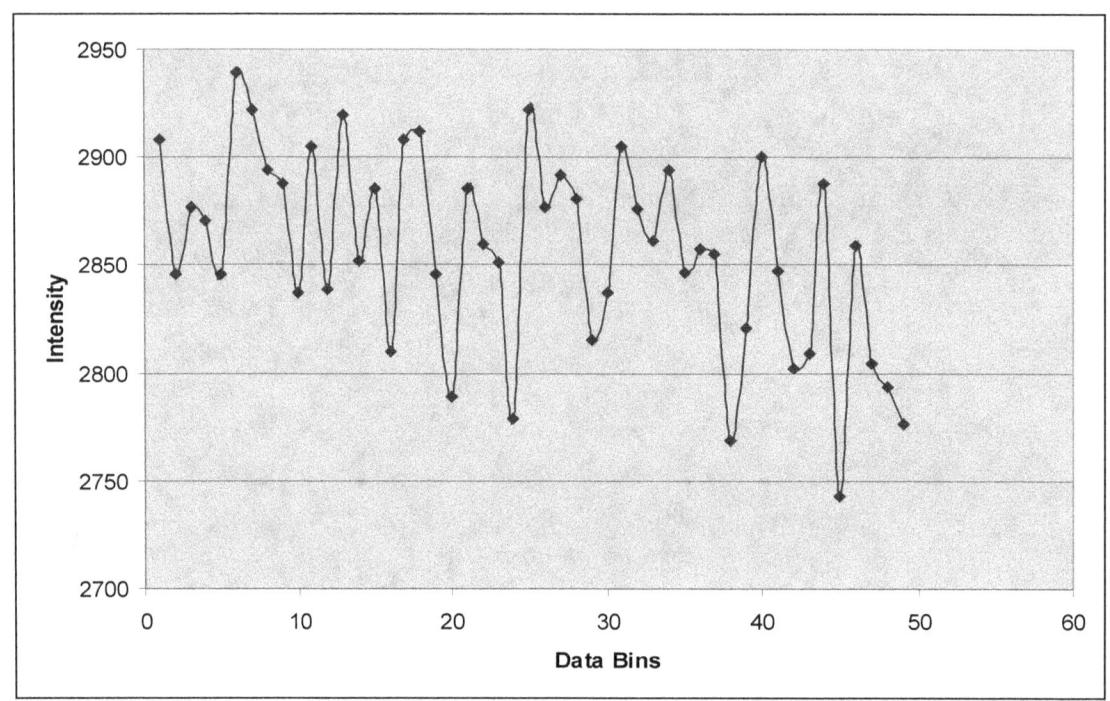

11. To change the graph of the data into a true spectrum, the x-axis must be changed to represent the frequency of the data point it is associated with. Looking at your raw data, note that column 10 is the first frequency channel and that column 11 is the change for each subsequent frequency.

    a. Make a new column for the frequency calculations.

    b. Do not do these calculations by hand! Use Excel to calculate the frequencies for you. First, input the value of the first frequency (i.e. the value in column 10) into the first cell of your new column.

    c. In the second cell of the column, type '="value from column 10" + "value from column 11'. Do not actually type the values of each respective column into the formula, rather, click on the values and they will automatically be used in the calculation.

        i. Since we are calculating the frequencies such that each subsequent frequency is equal to the previous frequency plus column 11, that value must be made a static, reference to remain the same in all calculations. To do this, simply place a $ sign in front of the letter and another $ in front of the number representing the value's local. For example, if the value is at K22, you would write it as K22 in the formula bar.

d. For the remaining cells, highlight the second cell and copy and paste it into the remaining cells.

e. Now, repeat the graphing procedure where the frequencies you calculated will be on the x-axis and the raw data on the y-axis. Be sure to label your axes. Your graph should look similar to this:

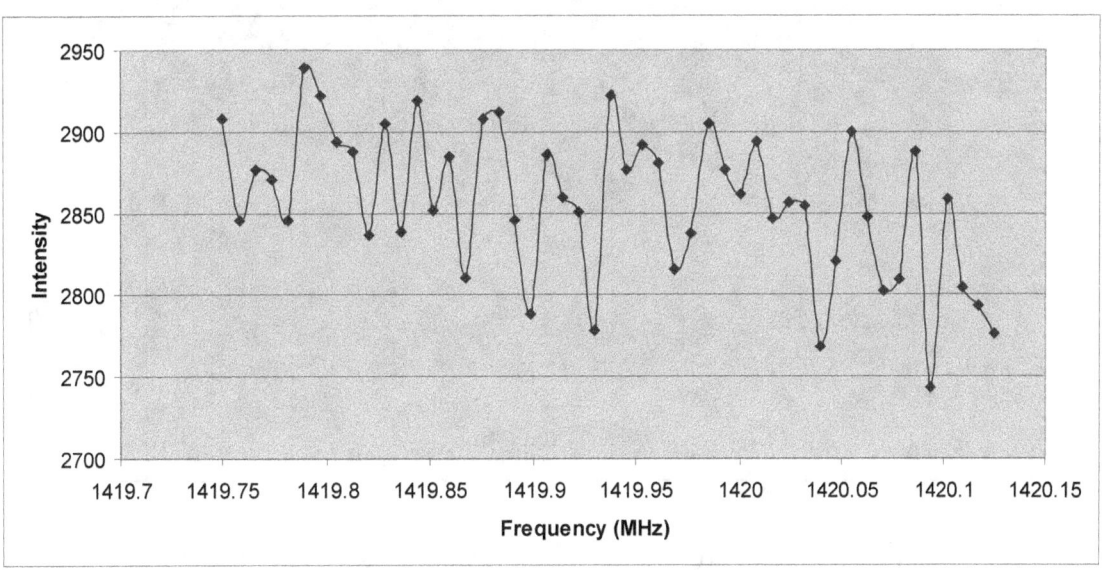

Part 2- Data Averaging

In part one, you used only one line of data to create a frequency plot. In the second section, you will average all your collected data and create a plot that represents all the data. This is to help you see the advantages of using more data points when analyzing data.

12. Start a new row below your data and use Excel's averaging formula, =AVG(data range) to create one averaged value for each column of raw data (for columns 14 and up). Now you should have just one row of data.

13. Now create a scatter plot with the averaged data on the y-axis and the number of columns on the x-axis. Your graph should look something like this:

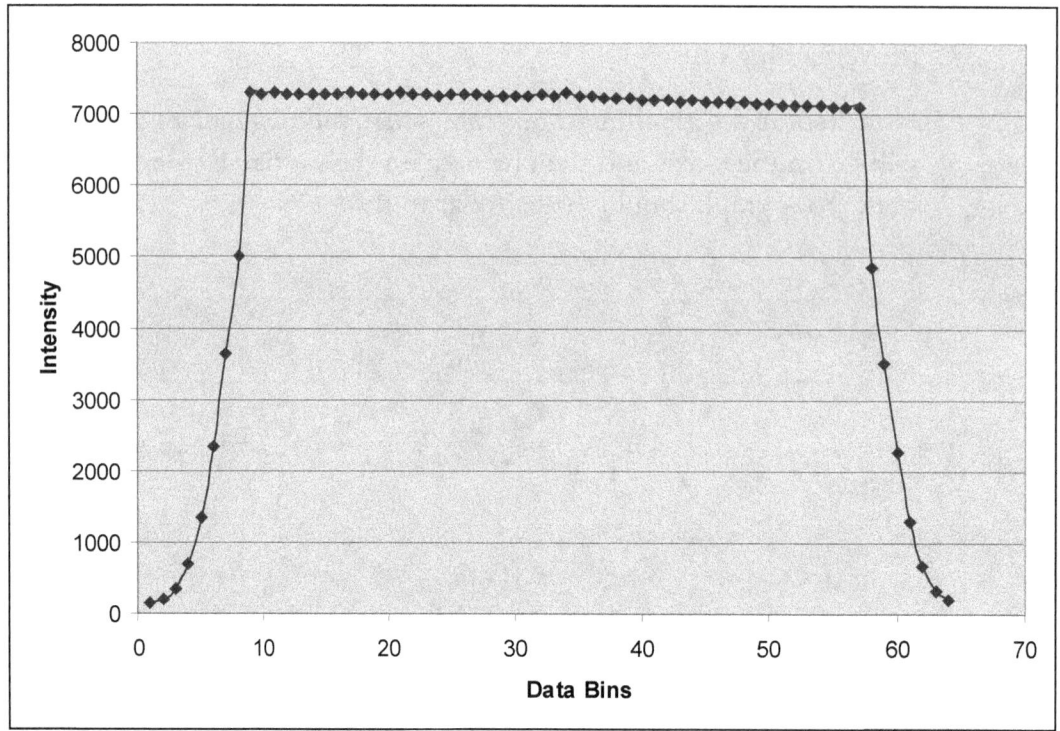

14. Repeat steps 9 and 10. Your graph should look something like this:

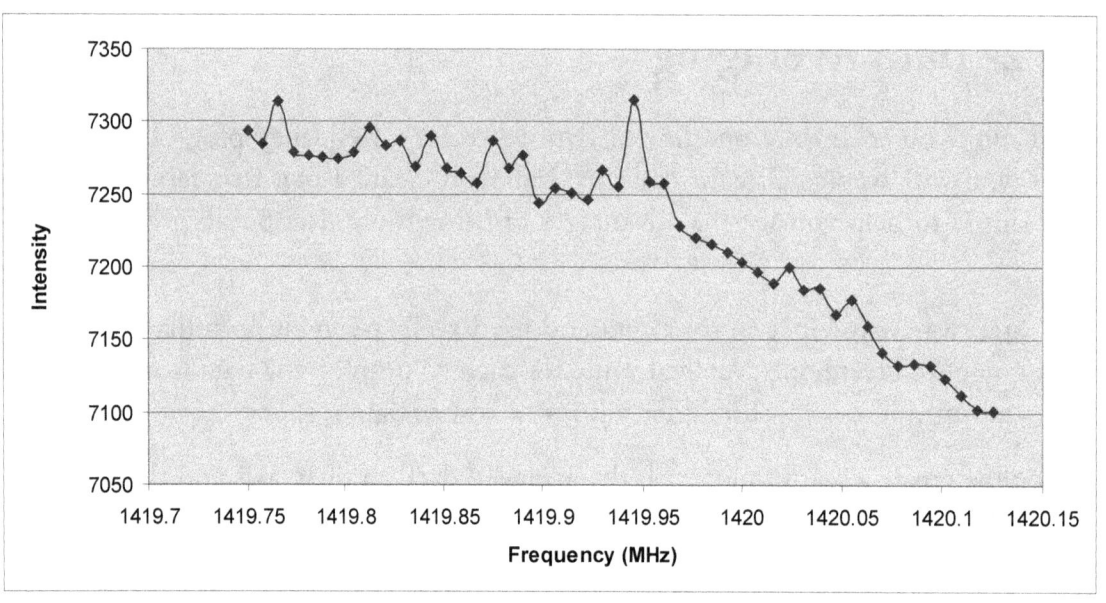

Questions

1. Compare the final graphs you got in Part 1 and Part 2. What differences do you notice in the graphs and why do you think there are such differences?

Lab 3: Intensity vs. Frequency Plot of Our Sun

Purpose: To create intensity versus frequency plots of the Sun in the SRT software and in Excel.

Procedure:

Plot in SRT

1. Open the SRT program

2. Calibrate the telescope (refer to Lab 1 if needed)

3. Click on the freq button to set the frequency and mode

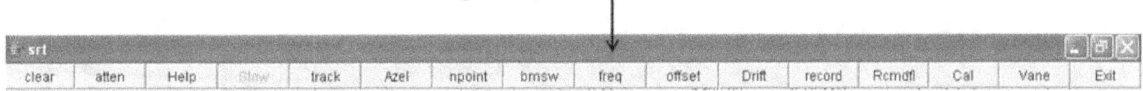

The following information will appear near the bottom left corner of the screen.

click to set center frequency (MHz), [optional mode]
mode 1 - default bandwidth = 500 kHz mode 2 - bw = 250 kHz mode 3 - bw = 125 kHz mode 4 - 3x500 kHz mode 5 - scanmode
the 21cm hydrogen line has a rest frequency of 1420.4 MHz
default frequency for continuum is 1420.0 MHz continuum uses average power from all frequencies

4. Type the number 1420 at the bottom left of the screen to set the telescope to look at frequencies centered around 1420 MHz. Do not hit return at this point.

5. Notice the mode option discussed in the text at the bottom left of the screen. This sets the bandwidth, i.e, the range of frequencies the telescope looks at. For example mode 1 has a bandwidth of 0.5 MHz, mode 2 has a bandwidth of 0.25 MHz, mode 3 has a bandwidth of 0.125 MHz, and mode 4 has a bandwidth of 3x.5 MHz.

6. Set the mode skipping a space after the frequency you typed in and typing 1. Your data input line should appear as "1420 1". This sets the telescope to look at the central frequency of 1420 MHz and a bandwidth of 0.5 MHz. Hit return.

7. Point the telescope at the Sun

8. Begin recording data (One minute of data recording is sufficient)

Note the red and black graphs at the top of the screen. The black graph shows the most recent data taken by the telescope. The red graph shows an averaging of the data over time.

9. Click on the red graph. It should appear in its own window.

10. Look at the graph and make sure the central frequency is 1420 MHz with a bandwidth of 0.5 MHz.

11. Make a copy of the graph to turn in by pressing the "printscreen" button on the upper right side of the keyboard.

12. Press the record button again to stop recording.

13. To display and print a copy of your graph, go to 'Start' on the Microsoft toolbar. Click on 'All programs', go to 'Accessories', and click on 'Paint'

14. Once Paint is open, you can either press Ctrl+V (the paste shortcut) or right-click the mouse and select the paste option. Your graph should now be displayed.

15. Stow the telescope and exit the SRT program.

Plot in Excel

16. Open the data you just created in Excel and reduce the data exactly as done in Lab 2.

Questions

1. Both the red and black graphs are showing data with a central frequency of 1420 MHz and a bandwidth of 0.5 MHz. Why is the red graph the one used?

2. How can you tell the bandwidth shown is 0.5 MHz from the graph produced in the SRT interface?

3. Set the telescope to a different central frequency and bandwidth. Predict the range of the x-axis you will see on your graph. Verify it.

Lab 4: Contour Map of the Sun

Purpose: To use the SRT software to create a 2-dimensional plot of the radio Sun.

Procedure:

1. Start the SRT program

2. Calibrate the telescope

3. Point the telescope at the Sun

4. Press the npoint button. The telescope will take measurements from 25 different points in the sky near and on the Sun. This will create a 25-pixel image of the area of the sky containing the Sun.

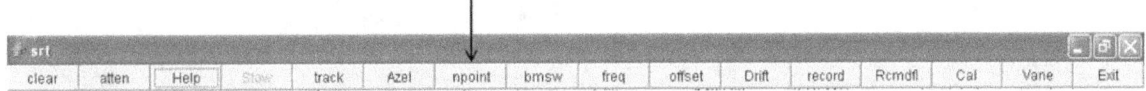

5. Once the telescope is done taking measurements, a red graph will appear in the upper-left part of the screen. *Note: If you minimize the SRT program or open another application once the npoint plot is displayed, the graph will disappear and you will have to start over*

6. The graph should look like the one below with the 'bull's eye'.

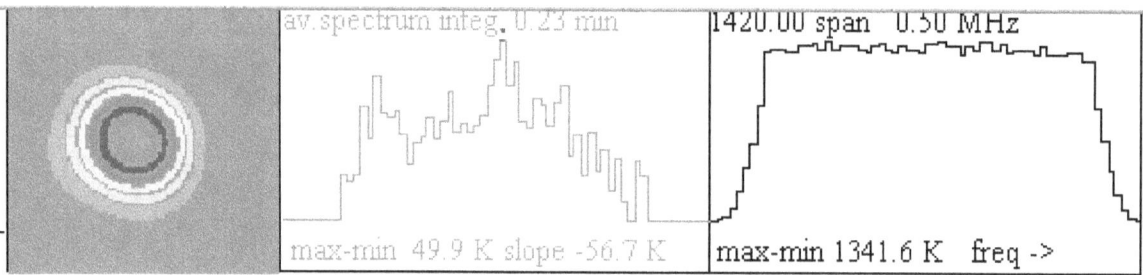

7. To get a copy of the graph use the printscreen button

8. Stow the telescope and exit the program

Questions:

1. Is your image of the Sun centered in your graph? If so, what does this tell you? If not, what does this tell you?

2. Describe in your own words how the bull's eye image is generated from the data scans taken by the telescope.

3. Would you be able to use the SRT to make npoint observations of any other source in the sky? Why or why not?

Lab 5: Measurement of Antenna Beamwidth

Objective: To measure the beamwidth of the SRT and compare with the diffraction limit for the telescope.

Introduction:

The SRT receives radio waves emitted by objects in the Universe, and measures them. For the telescope to detect the radio wave, it must be inside of "the beam", which means it arrives at the telescope dish and is reflected to the feed. Any radio waves which do not complete this process are considered "out of the beam". So knowing exactly "what the beam is" is important for understanding and analyzing your data.

When you point the SRT at the sky, it adds up all of the waves which enter it from a given direction into a single data point. It does not have the ability to create an image of what is in front of it at any given time. If you want to form an image with the SRT, you have to point and measure, then scan the telescope to point a few degrees away and measure, and repeat this procedure in both azimuth and elevation until the image is formed - - one pointing provides one pixel to the image. This is what you did in the npoint lab for our Sun.

To interpret the data we get from the telescope, we need to determine what is called the beamwidth. Technically, in radio astronomy the beamwidth of a radio telescope is defined as the "solid-angle measure of the half-power point of the main lobe of the antenna pattern". The half-power beamwidth (HPBW) can be measured by moving the telescope in a scan across a bright radio source. In this lab, you will determine the beamwidth of the SRT by scanning across the Sun.

There is a minimum value for the beamwidth which is determined by the wave nature of the light being detected and the physical parameters of the telescope reflector. This lower limit is called the diffraction limit and is given by

$$\theta = 1.22(\lambda/d),$$

where λ is the wavelength of the light being detected and d is the diameter of the aperture the light is entering (diameter of the telescope dish).

Many factors play a role in the actual beamwidth, however, and in this lab you will compare what you measure to this theoretical lower limit.

Procedure:

1. You will scan across the Sun in azimuth, starting 40° to the east of the Sun and going to 40° to the west of the Sun in 1° increments. To accomplish this task, you will need to make your measurements close to local noon. Make sure you calibrate the telescope on a part of the sky similar in elevation to the Sun. You may need to re-calibrate half way through the required observations.

2. Use a frequency of 1419 MHz. Record data for no more than 1 minute if the sky is clear.

3. Record the antenna temperature of the observation.

4. Record the azimuth and elevation of the Sun.

5. Change the azimuth of the observation by 1° and repeat until you have covered the necessary range.

6. Stow the telescope and exit the program.

Analysis:

1. Use your data to create a spreadsheet in Excel. One column should be Azimuth and the other should be Temperature.

2. Correct your azimuth angle for elevation (not all azimuth angles are created equal!). Create an "angle" column in he spreadsheet using the formula: angle = azimuth * cos(elevation).

3. Plot temperature vs angle, and print the graph (full page). Draw a curve of best fit for the data. (Do not connect-the-dots!)

4. Draw a horizontal line at the minimum temperature you measured.

5. Draw a horizontal line at the maximum temperature you measured.

6. Draw a horizontal line at the average of the maximum and minimum temperatures. This line represents the "half power" level.

7. Find the two angles where the temperature curve intersects the average temperature (or half power) line.

8. The difference between these two angles is the beamwidth of the SRT.

9. Calculate the diffraction limit of beamwidth for a telescope the same size as the SRT at 1419 MHz.

Questions:

1. How does the beamwidth you measured compare to the diffraction limit you can calculate?

2. What factors other than diffraction could affect the beamwidth of the SRT?

3. Could you measure the beamwidth of the SRT using something as a source *other* than the Sun? What requirements would there be for the source?

Lab 6: Measurement of the SRT Aperture Efficiency

Objective:

To determine the aperture efficiency of the SRT.

Introduction:

Aperture efficiency, η, is the ratio of the effective aperture of a radio telescope to the true aperture. The true aperture, A, is simply the geometric collecting area of the telescope surface (πr^2 in the case of the SRT). The effective aperture, A_e, is less than the true one because of such issues as part of the surface area being blocked by the support rods, part of the area being blocked by the feed, telescope surface irregularities, and so on. Mathematically, the efficiency is given by

$$\eta = A_e / A$$

It is not unusual for 30-50% of the power from the observed source to be lost before reaching the receiver. What is crucial, however, is to know exactly how much is lost so you can account for it in performing unit conversions between telescope raw counts and standard units such as Jy or sfu.

For the SRT, you can determine η from

$$\eta = 2 k T_a / F A$$

where T_a is the antenna temperature in K, k is Boltzmann's constant 1.38×10^{-23} W/Hz/K, F is the radio source flux density in janskys (1 Jy = 10^{-26} W/m²/Hz) , and A is the area of the reflector in m².

Several sources can be used to determine the efficiency of the SRT. The better ones include Cyg X-1 or Cas A. Given that both sources are fairly faint to the SRT, the procedures for either source are the same.

Procedure:

1. Move the telescope to an area about 10 degrees offset in azimuth from the location of the source you are using. The elevation should be about the same as the source you are using.

2. You will be making a continuum observation, so select a frequency away from the 1420.4 MHz hydrogen emission line. For example, you might set freq and mode to 1419 1.

3. Calibrate.

4. Move onto the source you will be using.

5. Observe for 300 seconds. Record the antenna temperature.

6. Stow the telescope and exit the program.

Analysis:

Look up F for the source you observed. Put the values into the equation for η.

Does your efficiency value make sense?

Observe another source and determine the value for η.

Do you get a similar value for η for both sources?

Lab 7: Measurement of the Sun's Flux Density

Objective:

Use the SRT to make measurements of the Sun's flux density to monitor solar activity such as flares.

Introduction:

Periodic observations of the Sun at a frequency of 1415.0 MHz using the SRT will allow a series of snapshots into the variability in radio emission generated by solar activity such as flares. Additional evidence of the flare will be gotten by correlating the SRT observations with observations at other wavelengths.

The radio emission at wavelengths from meters to tens of meters which comes from flares is quite different from emissions at shorter radio wavelengths. This has to do with the fact that at the shorter wavelengths the emission is caused by incoherent synchrotron radiation mechanisms while at the longer wavelengths the emission comes from coherent plasma radiation. The 1-3 GHz frequency range appears to lie at the transition between the two types of emission mechanisms. The shorter wavelength radio flares correlate well with X-ray events while the longer wavelength events appear not to be well correlated.

Solar flare emissions at X-ray wavelengths are sometimes accompanied by Coronal Mass Ejections (CMEs), which move away from the Sun at speeds exceeding 1000 km/s carrying strong magnetic fields and energetic particles. If Earth-directed, these CME events can deposit huge amounts of energy into Earth's upper atmosphere and trigger geomagnetic storms. These storms affect the atmospheric dynamics, leading to aurorae, satellite outages, power grid disruptions, and GPS navigation errors. So knowing when such flares occur is important, and the SRT provides a tool for detecting some of these solar flare events.

Procedure:

You will perform the observations with a central frequency of 1415.0 MHz, and a mode of 1. The data should be converted into an antenna temperature using the electronic noise calibration system on the SRT. The measurements should be made in a continuous mode from about 13:00 UT to about 20:00 UT.

You can obtain X-ray observations (1-min averages in the 0.1-0.8 nm waveband) of the Sun from the GOES-8 satellite at (http://www.sec.noaa.gov).

Analysis:

Create a series of plots for your observations. Read the data into Microsoft Excel, average, and plot. The easiest comparisons will be between raw (not calibrated) SRT data antenna temperature numbers and x-ray data that has been arbitrarily scaled so that they fit on the same plot for comparison purposes. A sample result is shown below:

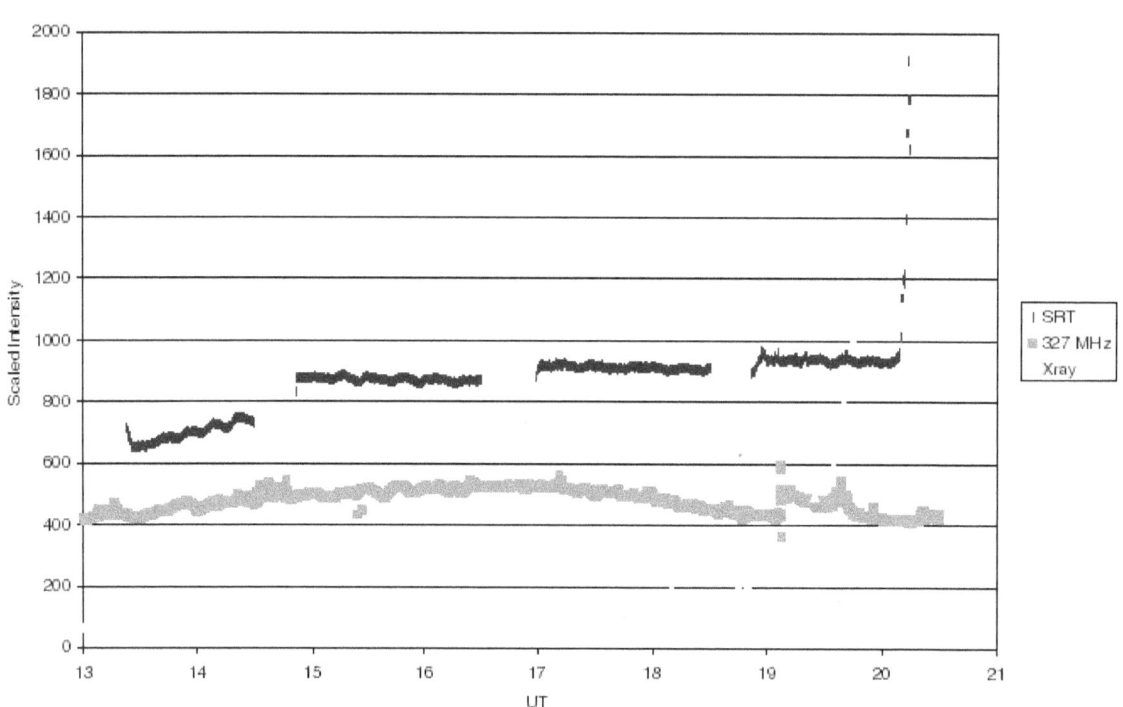

Graph courtesy of MIT Haystack Observatory

For each day, determine if an interesting event is seen in the SRT data. Is it correlated with the X-ray data? (You might consider doing a web search for other wavelength data sets to which you can correlate. Shown above is another set of observations at 327 MHz.)

Are there times when an event appears in one data set but not the other? How might you interpret this?

Lab 8: Solar Flux Density Variations and Solar Rotation

Objective:

A time history of flux density variations over an entire solar rotation might prove interesting and perhaps even provide evidence for the rotation period of the Sun. SRT data can then be compared to daily radio flux data from the NOAA site found at http://www.sec.noaa.gov/ftpdir/lists/radio/7day_rad.txt for additional insight.

It is interesting to note that the NIST radio station WWV out of Fort Collins, Colorado broadcasts the latest measurements of solar flux density in solar flux units (sfu) at 18 minutes past the hour, with an update every three hours, beginning at midnight UTC. You can also access the latest report at http://www.sec.noaa.gov/ftpdir/latest/wwv.txt. Data are gathered from a radio telescope in Ottawa, Canada.

Procedure:

1. Use a frequency of 1415 MHz with a mode of 1.

2. Convert the data to an antenna temperature using the noise calibration system on the SRT. Record the antenna temperature for each observation. Also record the system temperature for each observation.

3. Data should be taken between 13:00 UT and 20:00 UT. A 1-minute integration is fine under clear sky conditions.

4. Repeat your observations every day for a minimum of 40 days.

5. Stow the telescope when you have completed each measurement and exit the program.

Analysis:

1. Convert your antenna temperature data into solar flux data. To do this, you will need to know the SRT aperture efficiency. This value is determined in Lab 6, or you can repeat the measurement at this time to check for any changes in the value. The flux density F in units of janskys is equal to

$$F = (2 \, k \, T_a) \, / \, (\eta \, A \, 10^{-26})$$

where k is Boltzmann's constant 1.38×10^{-23} W/Hz/K, T_a is the antenna temperature in K, η is the aperture efficiency, and A is the geometric area of the dish on m^2.

2. Graph flux density versus day of observation. You might find it useful to convert the flux density units to solar flux units. One solar flux unit (sfu) is 10^{-22} W/m^2/Hz or 10,000 Jy.

3. Determine and remove any long-term trend from the data.

4. Calculate the error associated with each datum. You can determine the error associated with any T_a by using

$$\Delta T_a = T_s \, (Bt)^{-1/2}$$

where T_s is the system temperature, B is the instantaneous bandwidth for the observations, and t is the integration time over which the signal power is averaged.

5. Obtain data from NOAA for the same time period and scale it so you can plot it on the same graph as the SRT data. A sample graph for 28 days of data taken at the MIT Haystack SRT is shown below, with the lighter curve being the NOAA data and the darker curve (with error bars) being the SRT data.

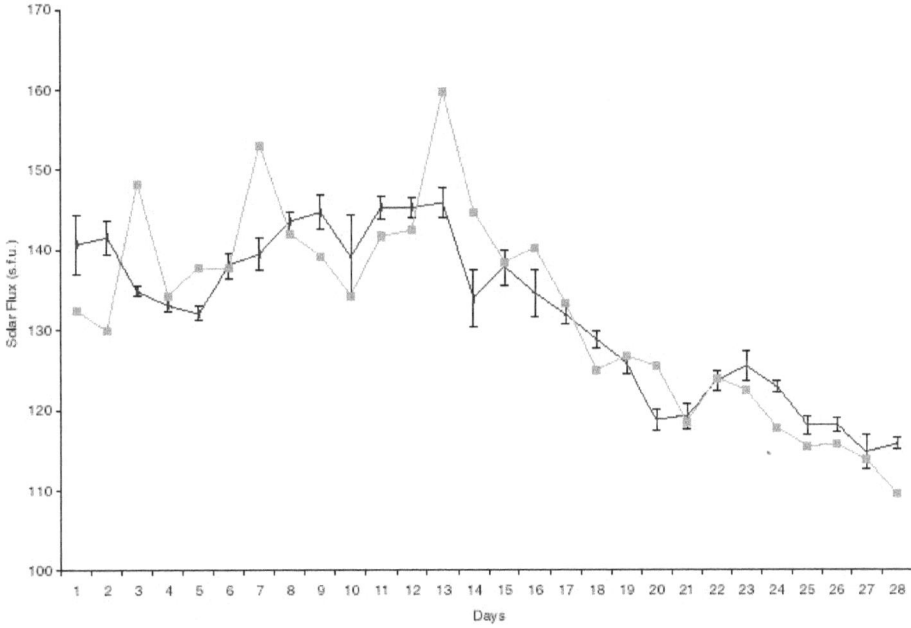

Graph courtesy of MIT Haystack Observatory

Questions:

1. Interpret your graph.

2. Based on your graph, what conclusions can you draw regarding the solar rotation period?

3. Compare your results to the known solar rotation period. How can you account for any differences?

Lab 9: Measurement of the Galactic Rotation Curve

Objective:

The purpose of this experiment is to create a rotational curve for the Milky Way Galaxy using 21-cm spectral line. The rotational curve will be created by plotting the maximum velocity observed along each line of sight versus the distance of this point from the Galactic center.

Introduction:

This lab is based upon using the Doppler shift to determine the velocity of a distant object. So how and why do you do what you do so as to answer the question? We must examine our galactic rotation model.

When we look at the luminous matter in distant galaxies, it is clear that most of it is in the center of the galaxy and it falls off exponentially as you move away from the center. This leads us to believe that we can model it as having all the mass in the center and the gravity felt by objects further out in the galaxy is simply given by Newton's Law of Gravitation,

$$F = G M m/r^2$$

where r is the distance from the galactic center to the orbiting object .This gravitational force must then provide the centripetal force

$$F = mv^2/r$$

to keep an object far out in the galaxy moving in an essentially circular orbit. Equating these two expressions, we see that the velocity of objects moving in circular orbits far from the centers of the galaxies should go as

$$v \sim r^{-1/2}.$$

So we would expect that if we graphed v vs r. we would find an exponential curve. If we find anything else, some explaining will have to take place.

The question then is what measurements do you make? We will determine the velocity of the emitting regions by using the Doppler shift, which tells us that the observed frequency (or wavelength) from an emitting object can be changed by the relative motion of that object to the observer. The line we will observe has a wavelength of 21-cm in the rest frame. Any shift from this particular wavelength is due to the Doppler shift and gives us a measure of the velocity of the emitting object (basically a cloud of hydrogen gas).

The Doppler shift is interested in the total relative velocity between emitter and observer. For the observer on Earth and the emitter a gas cloud in the Milky Way Galaxy, there are several motions to be considered. These include the motion of the Sun around the galactic center, the motion of the Earth around the Sun, the motion of the Earth around the common center of mass with the Moon, and the rotation of the Earth. The SRT software will take care of and remove the contributions of the motion of the Earth about the Sun, the motion of the Earth around its common center of mass with the Moon (usually too small to matter in this case), and the rotation of the Earth before it records any data. So it is only the motion of the Sun around the galactic center that we need to consider here.

Look at the diagram below. It is the radial velocity that is of interest to us. This is the component which causes the Doppler shift.

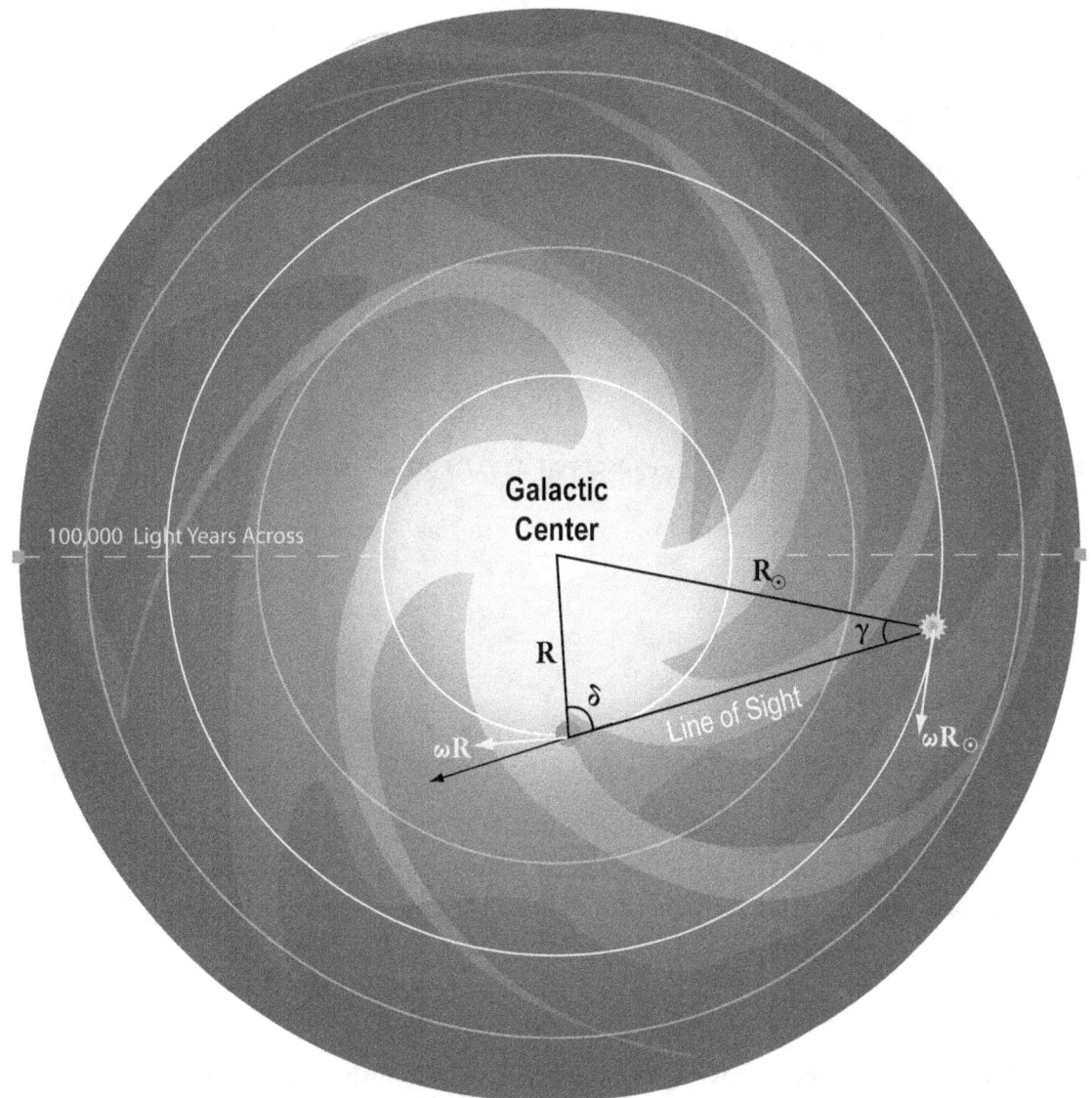

Recalling that the tangential velocity v is equal to the rotational velocity ω times the distance r the object is form the axis of rotation ($v = r\omega$), we see that if you are looking through the Galaxy at an angle γ from the center, the velocity of the gas at radius R projected along the line of site minus the velocity of the Sun projected on the same line is

$$V = \omega R \sin \delta - \omega_0 R_0 \sin \gamma .$$

where ω = angular velocity of source at distance R

ω_0 = angular velocity of Sun at a distance R_0

R_0 = distance of Sun to the Galactic center

γ = Galactic longitude of source

Using the trigonometric identity

$$\sin \delta = R_0 \sin \gamma / R$$

and substituting, we find that

$$V = (\omega - \omega_0) R_0 \sin \gamma$$

The maximum velocity occurs where the line of site is tangential to the circular motion in which case $\delta = 90°$ and we find that

$$R = \sin \gamma \, R_0$$

And hence V_{max} (at R) = ωR = V_{max} (observed at R) + $\omega_0 R_0 \sin \gamma$

From other measurements, we know that our Sun has a galactic rotational velocity and distance of

$$\omega_0 R_0 = 218 \text{ km/s}$$

$$R_0 = 2.6 \times 10^{17} \text{ km}$$

Now you have all of the background you need. V_{max} can be found using the Doppler shift.

You will make observations for various galactic longitudes. You will graph the observations for each coordinate location as a function of velocity, where the velocity is calculated from the Doppler shift equation

$$v = (\Delta\lambda/\lambda)c$$

where $\Delta\lambda$ is the difference between the observed wavelength and the known wavelength corresponding to the frequency of 1420.4 MHz, and c is the speed of light.

Clearly, the largest $\Delta\lambda$ corresponds to the cloud with the highest relative velocity (i.e., the one whose tangential velocity is along our line of sight) and given the wave equation $v = f\lambda$, we see that the largest wavelength corresponds to the smallest (or lowest frequency). This frequency (or frequency bin) is then identified in the data as the lowest frequency around 1420.4 MHz that has a non-zero intensity.

You are now ready to take your data and process it so that you can graph velocity versus radial distance from the galactic center. It will be helpful to use units of km/s for the velocity and kiloparsecs (kpc) for the distance. 1 kpc = 3.08568×10^{16} km.

Procedure:

1. Measurements should be made when the galactic center is within telescope limits.

2. Set the radiometer center frequency to be 1420.4 MHz. Set mode to be 4 which has a bandwidth of 1.5 MHz.

3. Set azel to be south at 45°, 180 45.

4. Calibrate.

5. Move telescope to galactic center, 0 0.

6. Begin recording data. Call the file g00.rad to stand for galactic longitude of 0° so you can know in the future which file is which data set.

7. Take 600 seconds of data.

8. Turn off recording.

9. The real time control program contains a routine that calculates the V_{LSR}. You must copy this data point down yourself – it is displayed in the lower right hand corner of the SRT software. This is the velocity of the local solar region (our approximate speed relative to the galaxy).

10. Move the telescope to a longitude of 10°, 10 0.

11. Record g10.rad.

12. REPEAT for longitudes every 10° up to and including 90°.

13. Stow the telescope and exit the program.

Analysis:

You now have 10 spectra. Read off the lowest frequency signal detected in each spectrum. Use the minimum frequency to calculate the maximum velocity associated with it. This is the maximum velocity emission relative to the local standard of rest (that is, after the Doppler shift introduced by the local motions of the Earth and Sun have been removed).

The velocity of the frequency channel is obtained by

$$V = ((1420.406-f) \, V_c / 1420.406) - V_{lsr}$$

where,

V_c = velocity of light; 299,790 km/s

V_{lsr} = velocity of observer relative to local standard of rest f = frequency of channel = $f_{L.O.} + 0.04$

Complete your calculations of $V_{max} = V_{max}$ observed + $\omega_0 R$, then graph V_{max} vs. Distance from Galactic Center.

Questions:

1. Describe what you see in your graph? Does one line or curve of best fit account for the entire data set? If so, what is the best fit? If not, at what distance does a break occur?

2. What Galactic mass distribution might fit your data? What galactic mass distribution does not fit your data?